What are ATOMS?

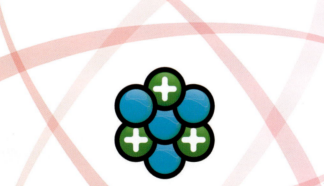

By Nathan Miloszewski

Gareth Stevens
PUBLISHING

Please visit our website, www.garethstevens.com. For a free color catalog of all our high-quality books, call toll free 1-800-542-2595 or fax 1-877-542-2596.

Library of Congress Cataloging-in-Publication Data

Names: Miloszewski, Nathan, author.
Title: What are atoms? / Nathan Miloszewski.
Description: New York : Gareth Stevens Publishing, [2022] | Series: The facts on matter | Includes index.
Identifiers: LCCN 2020038073 (print) | LCCN 2020038074 (ebook) | ISBN 9781538267011 (library binding) | ISBN 9781538266991 (paperback) | ISBN 9781538267004 (set) | ISBN 9781538267028 (ebook)
Subjects: LCSH: Atoms–Juvenile literature.
Classification: LCC QC173.16 .M55 2022 (print) | LCC QC173.16 (ebook) | DDC 539.7–dc23
LC record available at https://lccn.loc.gov/2020038073
LC ebook record available at https://lccn.loc.gov/2020038074

First Edition

Published in 2022 by
Gareth Stevens Publishing
111 East 14th Street, Suite 349
New York, NY 10003

Copyright © 2022 Gareth Stevens Publishing

Designer: Katelyn E. Reynolds
Editor: Char Light

Photo credits: Cover, pp. 1 (atom), 6, 9, 11 VectorMine/ iStock / Getty Images Plus; cover, pp. 1–24 (background) sudanmas/ E+/Getty Images; cover, pp. 1–24 (chemistry doodles) backUp/Shutterstock.com; cover, pp. 1–24 (banner) Ozz Design/ Shutterstock.com; cover, pp. 1–24 (paper) Sergey Mironov /Shutterstock.com; p. 5 Roquillo Tebar/Shutterstock.com; p. 5 StockImageFactory.com/Shutterstock.com; p. 5 lunamarina/Shutterstock.com; p. 7 blueringmedia/ iStock / Getty Images Plus; p. 7 South_agency/ E+/Getty Images; p. 10 courtesy of the Library of Congress; pp. 13, 19 duntaro/ iStock / Getty Images Plus; p. 13 JGI/Jamie Grill/Getty Images; p. 15 LuFeeTheBear/Shutterstock.com; p. 15 Maksim Shmeljov/Shutterstock.com; p. 17 avid_creative/E+/Getty Images; p. 21 ANDREW CABALLERO-REYNOLDS/AFP via Getty Images.

Printed in the United States of America

Some of the images in this book illustrate individuals who are models. The depictions do not imply actual situations or events.

CPSIA compliance information: Batch #CWGS22: For further information contact Gareth Stevens, New York, New York at 1-800-542-2595.

Find us on

CONTENTS

Words in the glossary appear in **bold** type the first time they are used in the text.

Atoms Make Up
EVERYTHING

Have you ever built a castle out of plastic bricks, like LEGOs? The castle is made up of many smaller bricks. Everything around you works the same way—the bed you sleep in, a tree you climb, and even your own body! All of these things are made up of tiny "building blocks" called atoms.

An atom is the smallest particle, or tiny piece, of an element that still has its special properties. All living things and all matter in the **universe** are made of atoms!

Know the Basics!

Elements are made of only one kind of atom. Your body is made up mostly of the elements oxygen, hydrogen, carbon, and nitrogen.

You, and everything around you, are made up of many, many atoms!

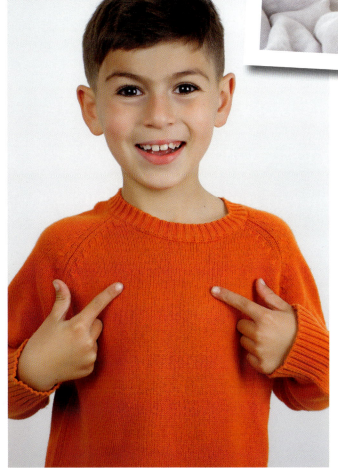

Using Blocks to
BUILD THINGS

When you are playing with your plastic bricks, all of those

small pieces come together to build something bigger—like

a house.

An atom is like a single brick or block. When two or more

atoms are joined, or bonded, they form a molecule—just like

when you click plastic bricks together to build a wall or building.

If you look closely enough,

you'll see that all solids,

liquids, and gases are

made of many molecules.

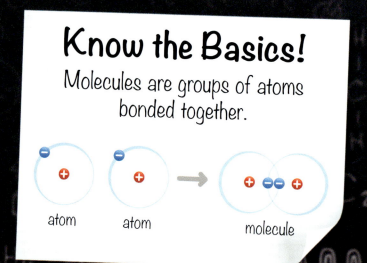

Know the Basics!
Molecules are groups of atoms bonded together.

atom   atom   molecule

6

gas

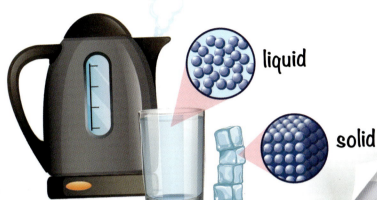
liquid

solid

The three main states of matter are solid, liquid, and gas. Molecules are packed together tightly in a solid. They move around more in liquids and gases.

particles. "Sub" means below. So, subatomic means below the level of an atom, or smaller than an atom.

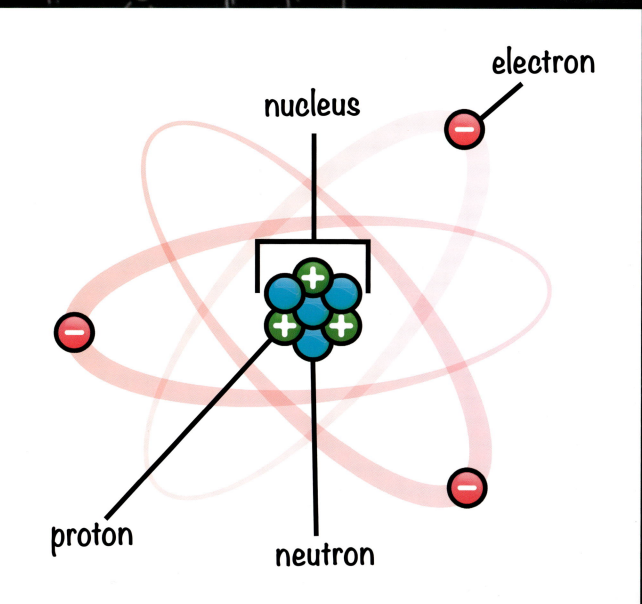

electron

nucleus

proton

neutron

What Is a PROTON?

Protons can be found in all atoms. They have a positive electric **charge**. They make up half of the atom's nucleus.

Knowing how many protons are in an atom's nucleus is important. It can tell you what kind of atom it is. Then you'll know what element it is too. Elements have special features, or properties. The number of protons in an atom's nucleus also gives an element its atomic number.

Know the Basics!

On the **periodic table of elements**, the elements are arranged by their atomic number.

atomic number ——— 2

He

symbol

name ——— Helium

4.0026 ——— atomic weight

"He" stands for the element helium. It's a gas often used to fill balloons! The number 2 is its atomic number. That means it has two protons in its nucleus.

What Is a NEUTRON?

Neutrons are neutral, which means they have no electric charge. They're the largest particles inside of an atom. They're the other half of the nucleus in the center of an atom.

The nucleus makes up 99.99% of the mass of an atom. As some elements grow older, their atoms lose neutrons. This breaking down and falling apart is called radioactive decay. Radioactive decay helps scientists figure out how old something is, like dinosaur bones.

Know the Basics!

When living things decay, they leave behind the element carbon.

Knowing how long it takes for carbon to break down helps scientists figure out the age of these dinosaur bones. This is called carbon dating.

What Is an ELECTRON?

Electrons are particles with a **negative** charge. They spin very fast around the nucleus of an atom. They're so small—1,800 times smaller than protons and neutrons—that their mass is almost zero. They have lots of energy, or power, so they are constantly moving.

Electrons help bond atoms together. Atoms bonding is what causes **chemical reactions**!

Know the Basics!

Metal can **conduct** electricity, which is a current, or stream, of electrons through the metal.

Atoms are so small that you must use
a very powerful kind of **microscope** called
an electron microscope to see them.

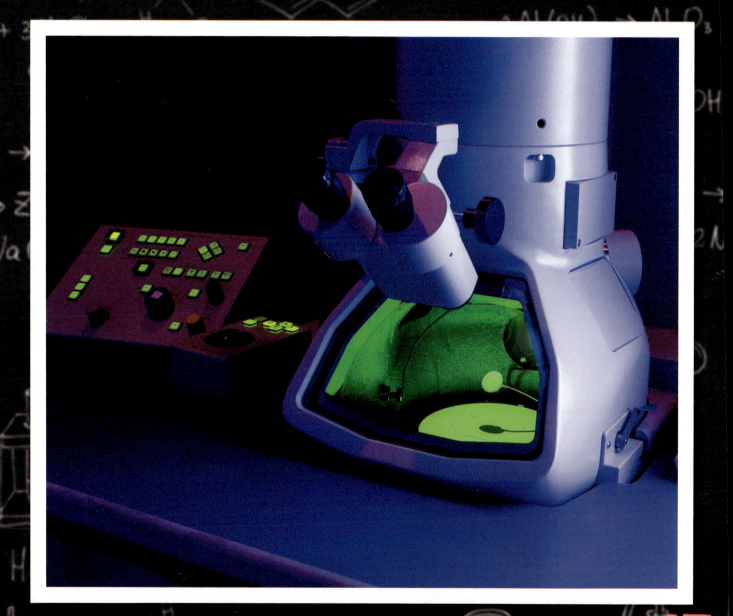

How Are Atoms Held TOGETHER?

The glue that holds molecules together is called a chemical bond. There are three main types:

- **Ionic bond:** This "addition and subtraction bond" joins the atom of a metal to a **non-metal**. The metal atom loses an electron, and the non-metal atom gains one.

- **Covalent bond:** Non-metal atoms share electrons in this kind of bond.

- **Metallic bond:** The "heat and electricity bond" holds metal atoms together. This bond is what allows people to use heat for cooking or electricity.

THE 3 MAIN TYPES OF CHEMICAL BONDS

IONIC BOND

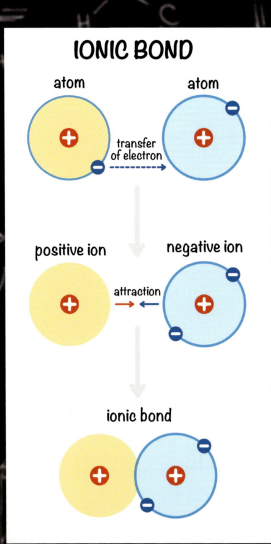

atom   atom

transfer of electron

positive ion   negative ion

attraction

ionic bond

COVALENT BOND

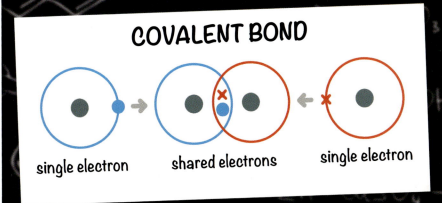

single electron   shared electrons   single electron

METALLIC BOND

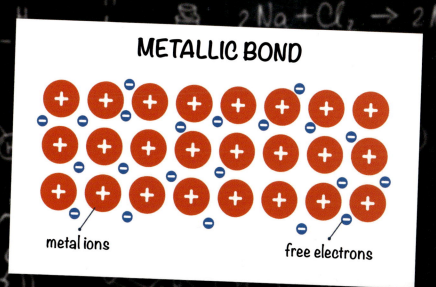

metal ions   free electrons

When atoms bump into each other, a chemical bond is what makes them stick together. There are different types of bonds between different types of matter.

The Elements and Their
PERIODIC TABLE

The periodic table of elements is a list of 118 elements that scientists currently know about. The three main types of elements are metals, non-metals, and **metalloids**. Most of the elements are metals—things that are shiny and can conduct electricity. Elements are mostly solids. Some are gases, and even fewer are liquids.

The table of elements arranges the elements by their atomic number from the lowest to the highest. In 2016, four new elements were added to the table! Will there be more?

Know the Basics!

You already know one important element! Oxygen is the gas humans and animals need to breathe. Its atomic number is 8.

PERIODIC TABLE OF ELEMENTS

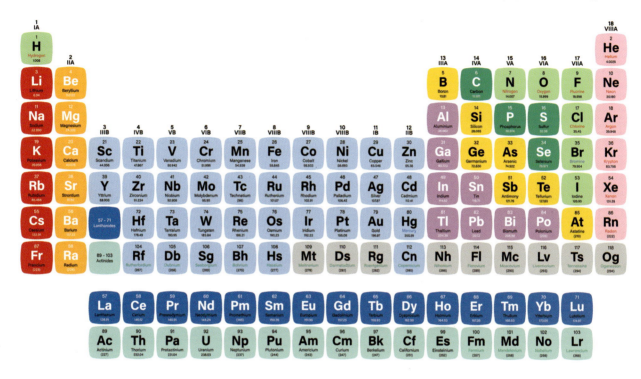

1 IA																	18 VIIIA
1 **H** Hydrogen 1.008	2 IIA											13 IIIA	14 IVA	15 VA	16 VIA	17 VIIA	2 **He** Helium 4.0026
3 **Li** Lithium 6.94	4 **Be** Beryllium 9.0122											5 **B** Boron 10.81	6 **C** Carbon 12.011	7 **N** Nitrogen 14.007	8 **O** Oxygen 15.999	9 **F** Fluorine 18.998	10 **Ne** Neon 20.180
11 **Na** Sodium 22.990	12 **Mg** Magnesium 24.305	3 IIIB	4 IVB	5 VB	6 VIB	7 VIIB	8 VIIIB	9 VIIIB	10 VIIIB	11 IB	12 IIB	13 **Al** Aluminium 26.982	14 **Si** Silicon 28.085	15 **P** Phosphorus 30.974	16 **S** Sulfur 32.06	17 **Cl** Chlorine 35.45	18 **Ar** Argon 39.948
19 **K** Potassium 39.098	20 **Ca** Calcium 40.078	21 **Sc** Scandium 44.956	22 **Ti** Titanium 47.867	23 **V** Vanadium 50.942	24 **Cr** Chromium 51.996	25 **Mn** Manganese 54.938	26 **Fe** Iron 55.845	27 **Co** Cobalt 58.933	28 **Ni** Nickel 58.693	29 **Cu** Copper 63.546	30 **Zn** Zinc 65.38	31 **Ga** Gallium 69.723	32 **Ge** Germanium 72.630	33 **As** Arsenic 74.922	34 **Se** Selenium 78.971	35 **Br** Bromine 79.904	36 **Kr** Krypton 83.798
37 **Rb** Rubidium 85.468	38 **Sr** Strontium 87.62	39 **Y** Yttrium 88.906	40 **Zr** Zirconium 91.224	41 **Nb** Niobium 92.906	42 **Mo** Molybdenum 95.95	43 **Tc** Technetium (98)	44 **Ru** Ruthenium 101.07	45 **Rh** Rhodium 102.91	46 **Pd** Palladium 106.42	47 **Ag** Silver 107.87	48 **Cd** Cadmium 112.41	49 **In** Indium 114.82	50 **Sn** Tin 118.71	51 **Sb** Antimony 121.76	52 **Te** Tellurium 127.60	53 **I** Iodine 126.90	54 **Xe** Xenon 131.29
55 **Cs** Caesium 132.91	56 **Ba** Barium 137.33	57 - 71 Lanthanides	72 **Hf** Hafnium 178.49	73 **Ta** Tantalum 180.95	74 **W** Tungsten 183.84	75 **Re** Rhenium 186.21	76 **Os** Osmium 190.23	77 **Ir** Iridium 192.22	78 **Pt** Platinum 195.08	79 **Au** Gold 196.97	80 **Hg** Mercury 200.59	81 **Tl** Thallium 204.38	82 **Pb** Lead 207.2	83 **Bi** Bismuth 208.98	84 **Po** Polonium (209)	85 **At** Astatine (210)	86 **Rn** Radon (222)
87 **Fr** Francium (223)	88 **Ra** Radium (226)	89 - 103 Actinides	104 **Rf** Rutherfordium (267)	105 **Db** Dubnium (268)	106 **Sg** Seaborgium (269)	107 **Bh** Bohrium (270)	108 **Hs** Hassium (277)	109 **Mt** Meitnerium (278)	110 **Ds** Darmstadtium (281)	111 **Rg** Roentgenium (282)	112 **Cn** Copernicium (285)	113 **Nh** Nihonium (286)	114 **Fl** Flerovium (289)	115 **Mc** Moscovium (290)	116 **Lv** Livermorium (293)	117 **Ts** Tennessine (294)	118 **Og** Oganesson (294)

57 **La** Lanthanum 138.91	58 **Ce** Cerium 140.12	59 **Pr** Praseodymium 140.91	60 **Nd** Neodymium 144.24	61 **Pm** Promethium (145)	62 **Sm** Samarium 150.36	63 **Eu** Europium 151.96	64 **Gd** Gadolinium 157.25	65 **Tb** Terbium 158.93	66 **Dy** Dysprosium 162.50	67 **Ho** Holmium 164.93	68 **Er** Erbium 167.26	69 **Tm** Thulium 168.93	70 **Yb** Ytterbium 173.05	71 **Lu** Lutetium 174.97
89 **Ac** Actinium (227)	90 **Th** Thorium 232.04	91 **Pa** Protactinium 231.04	92 **U** Uranium 238.03	93 **Np** Neptunium (237)	94 **Pu** Plutonium (244)	95 **Am** Americium (243)	96 **Cm** Curium (247)	97 **Bk** Berkelium (247)	98 **Cf** Californium (251)	99 **Es** Einsteinium (252)	100 **Fm** Fermium (257)	101 **Md** Mendelevium (258)	102 **No** Nobelium (259)	103 **Lr** Lawrencium (266)

1
H
Hydrogen
1.008

118
Og
Oganesson
(294)

Hydrogen has the lowest atomic number of 1. Oganesson has the highest atomic number of 118.

19

Energy and Life FROM ATOMS

We are made of atoms and so are our friends and family! Atoms make up the water we drink, the air we breathe, the food we eat, and the supplies we use to build our homes.

Atoms also change in natural ways and can be changed by humans to help make **technology** better. When split apart, atoms can release, or give off, energy we can use to power whole cities! Atoms may be small, but they matter a lot!

Nuclear power plants use nuclear fission, or the splitting of atoms, to make power.

GLOSSARY

charge: an amount of electricity

chemical reaction: the process that turns one substance into another

conduct: the ability to allow heat or electricity to move from one place to another

metalloid: an element that has a mix of properties from both metals and non-metals

microscope: a tool used to view very small objects so they can be seen much larger and more clearly

model: a set of ideas that describes something

negative: electricity that is charged by an electron

non-metal: something that is not a metal

orbit: a path that one object takes around another

periodic table of elements: a list of the elements scientists know about

technology: using science, engineering, and other industries to invent useful tools or to solve problems

universe: everything that exists

For More
INFORMATION

Books

Gray, Theodore. *The Kid's Book of the Elements: An Awesome Introduction to Every Known Atom in the Universe*. New York, NY: Black Dog & Leventhal, 2020.

National Geographic Kids. *Science Encyclopedia: Atom Smashing, Food Chemistry, Animals, Space, and More!* Washington, DC: National Geographic, 2016.

Zovinka PhD, Edward P. *Real Chemistry Experiments: 40 Exciting STEAM Activities for Kids*. Emeryville, CA : Rockridge Press, 2019.

Websites

Britannica Kids - Atoms
kids.britannica.com/kids/article/atom/352802
An introduction to atoms, their structure, and properties.

Energy Source Fact Files!
www.funkidslive.com/learn/energy-sources/uranium-nuclear-power-energy-source-fact-file-2/#
See how uranium atoms are split to release energy in a process called nuclear fission.

PBS - DK Find Out - What Is Matter?
www.dkfindout.com/us/science/solids-liquids-and-gases/what-is-matter/
Learn more about matter at this interactive website for kids.

INDEX